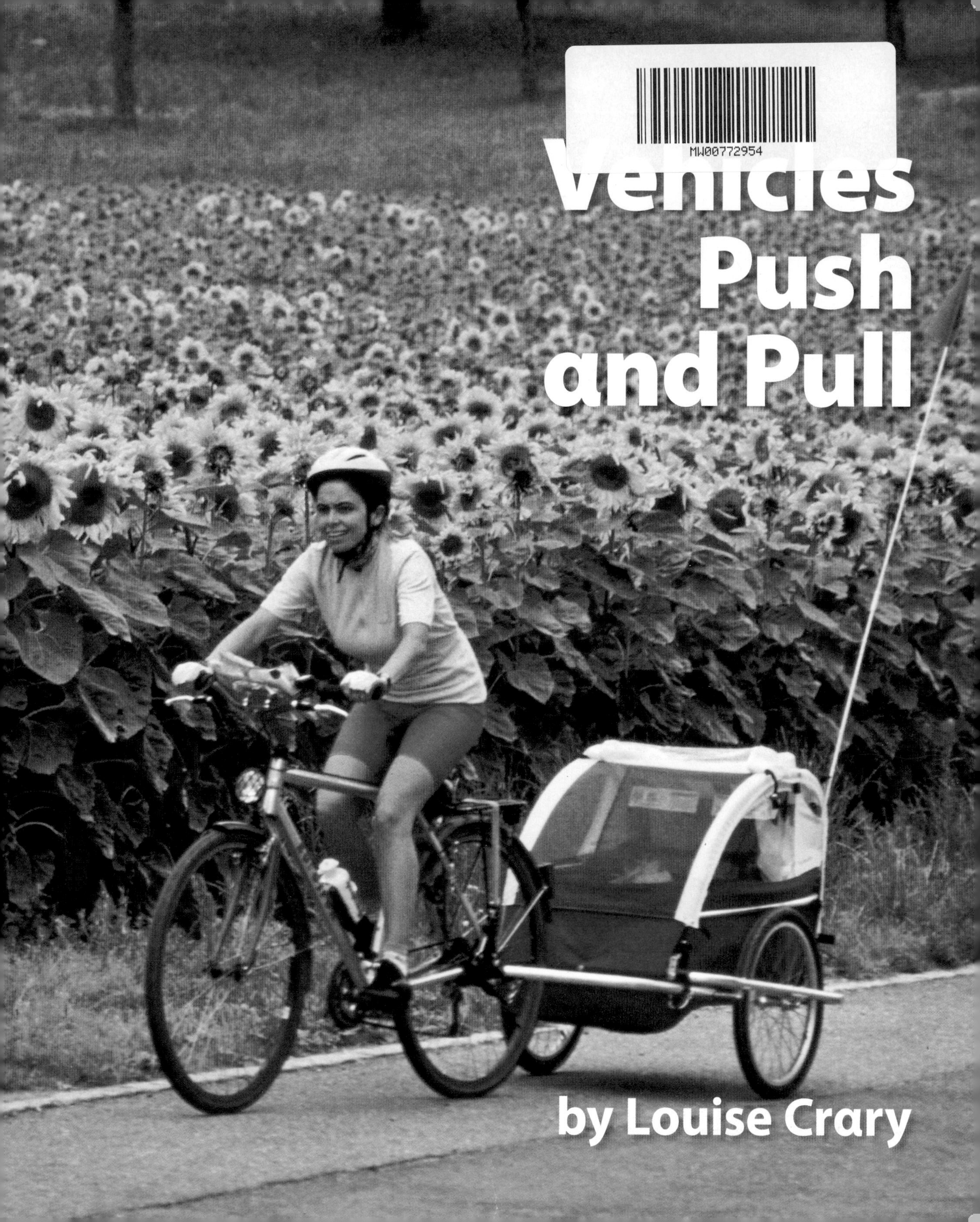

Vehicles Push and Pull

by Louise Crary

Vehicles **push**.

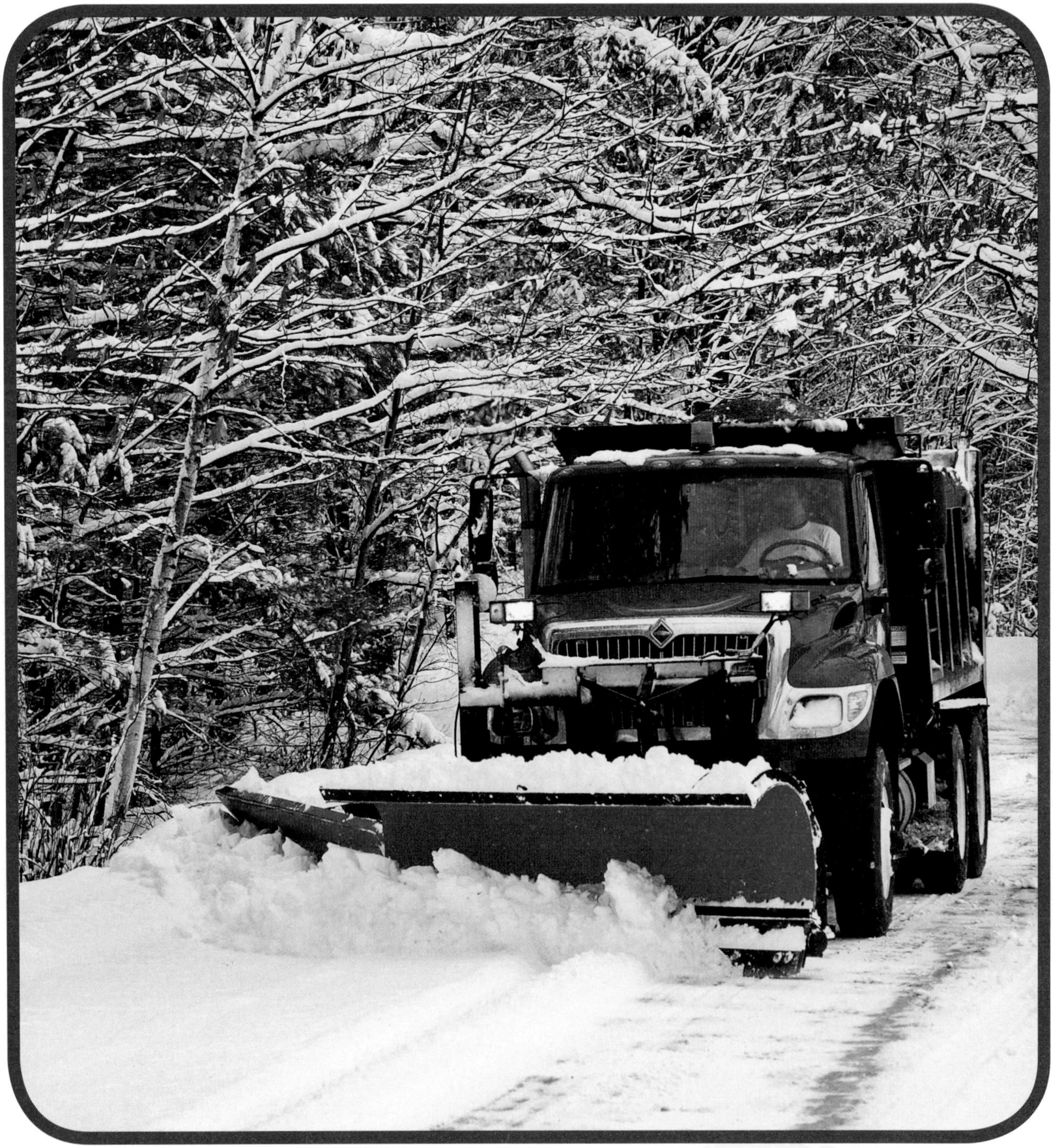

Vehicles **pull**.

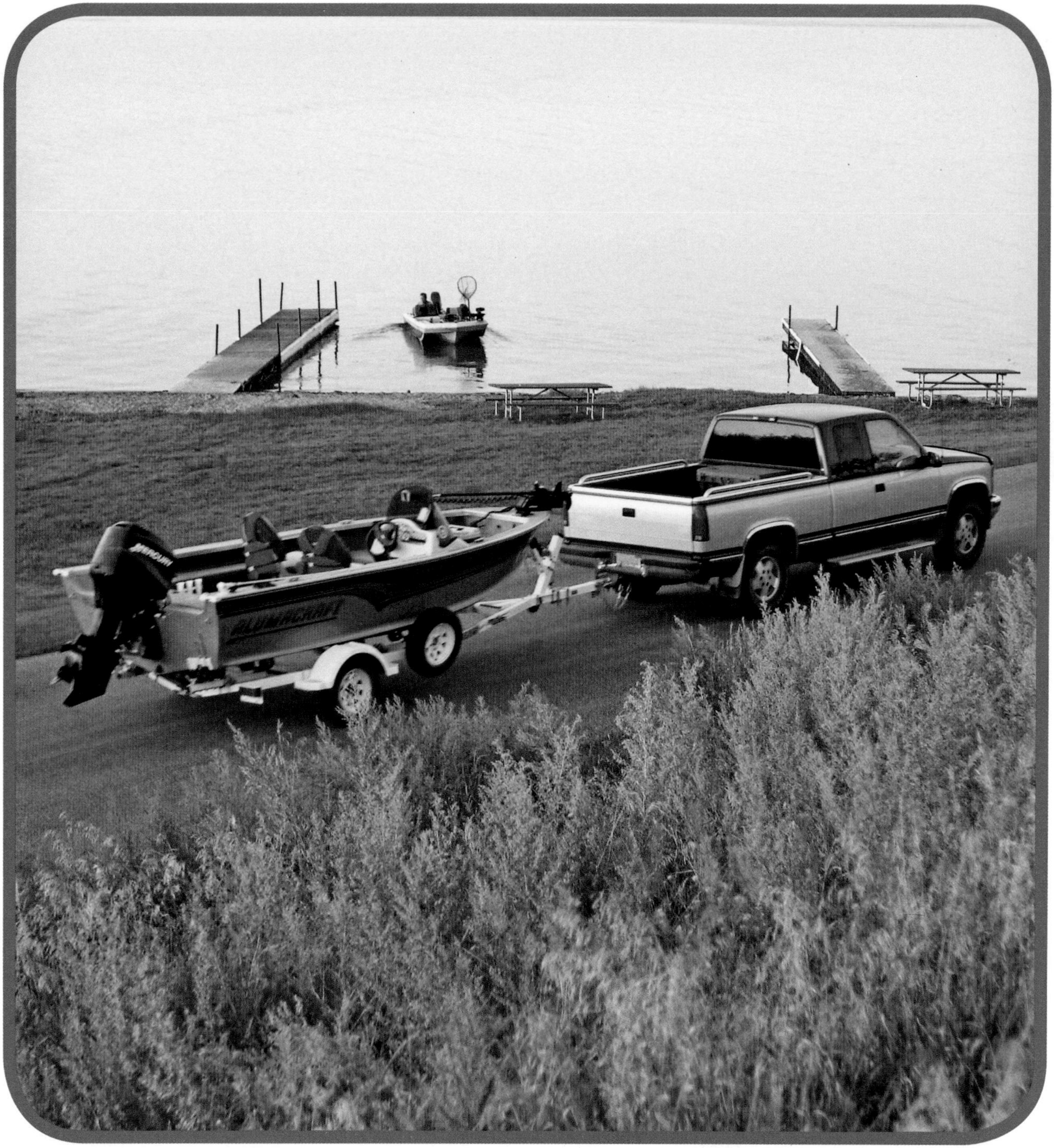

Vehicles push.

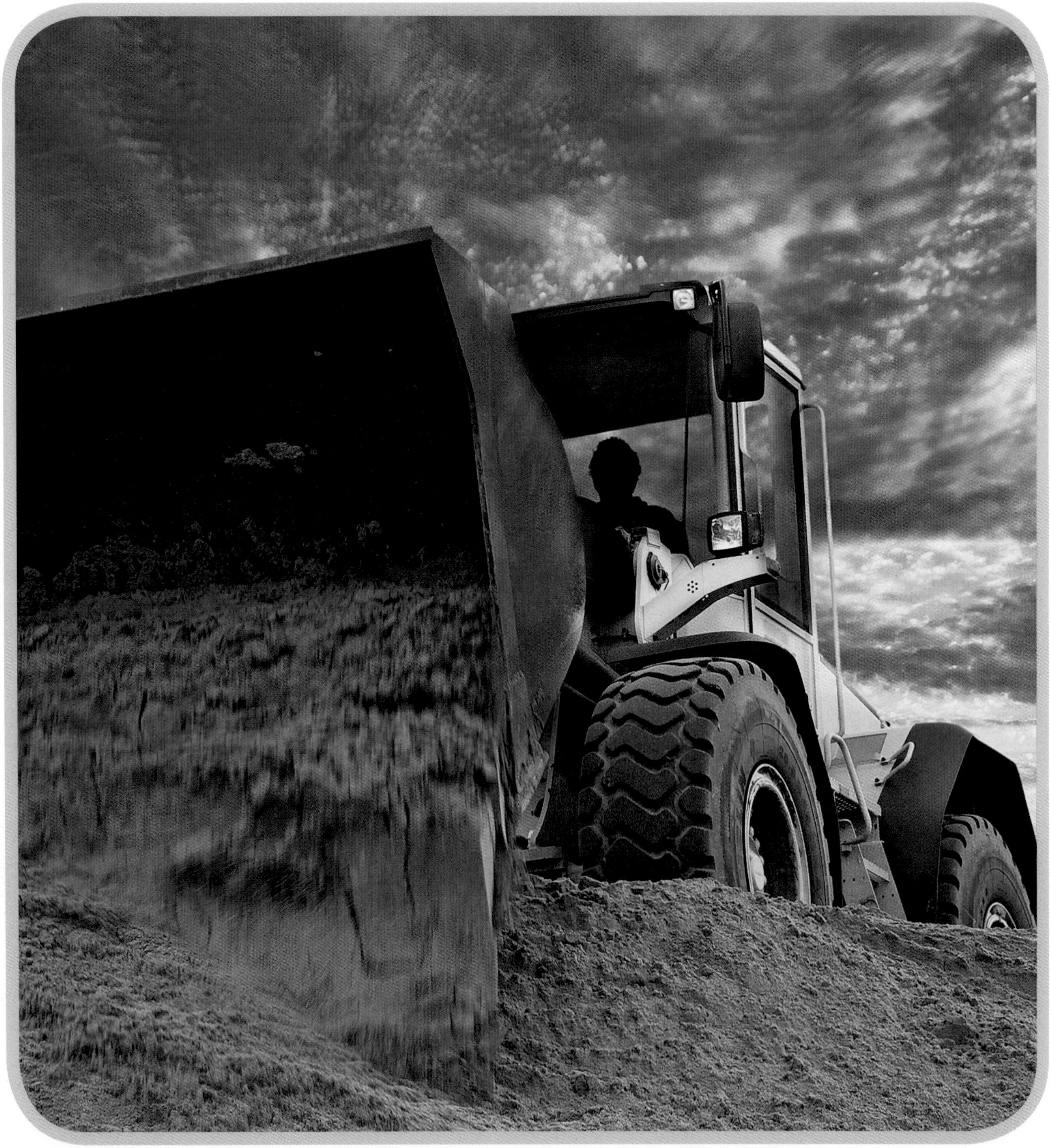

Vehicles pull.

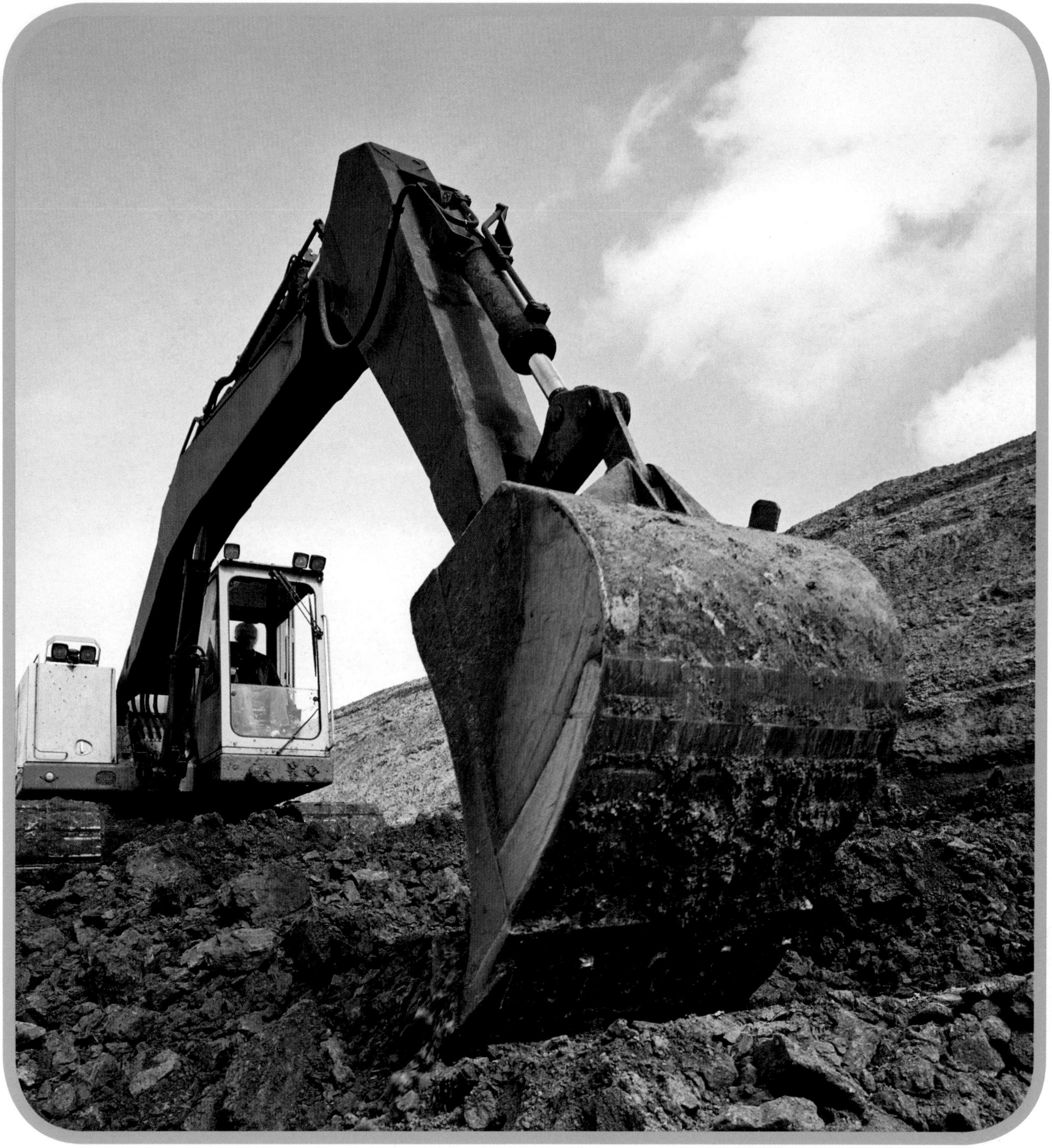

Vehicles push.

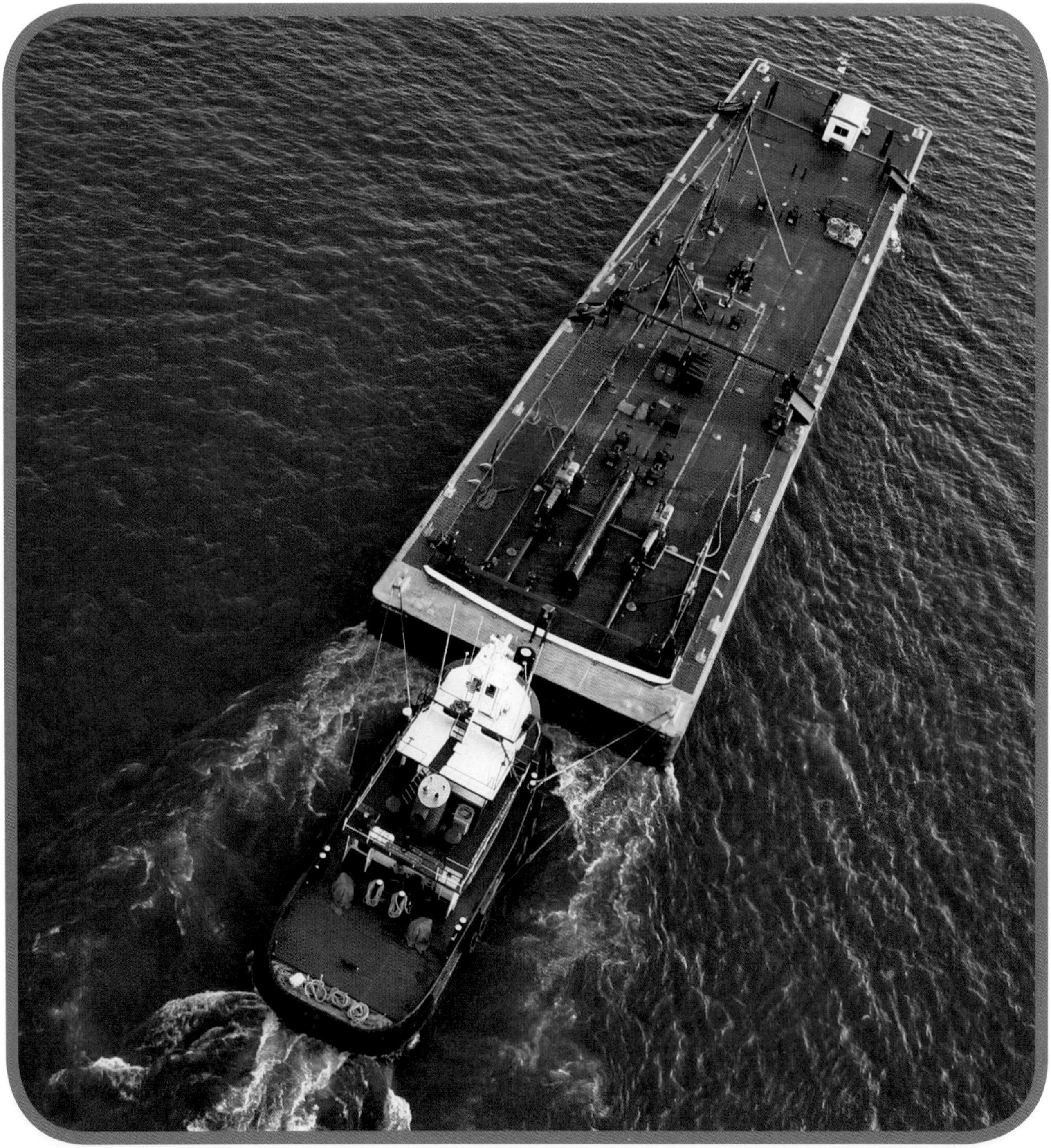

Vehicles pull.

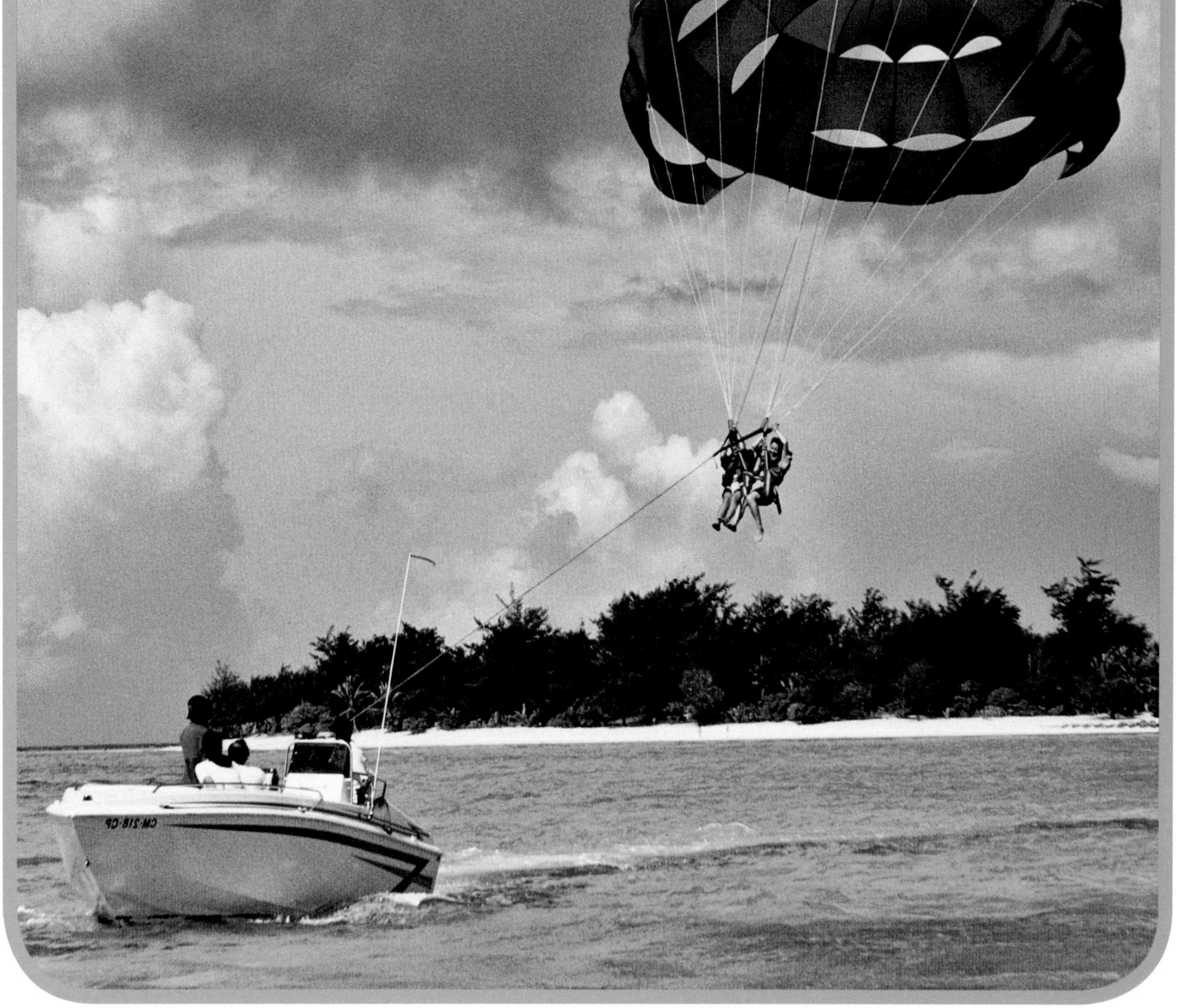

Vehicles Push and Pull

Push

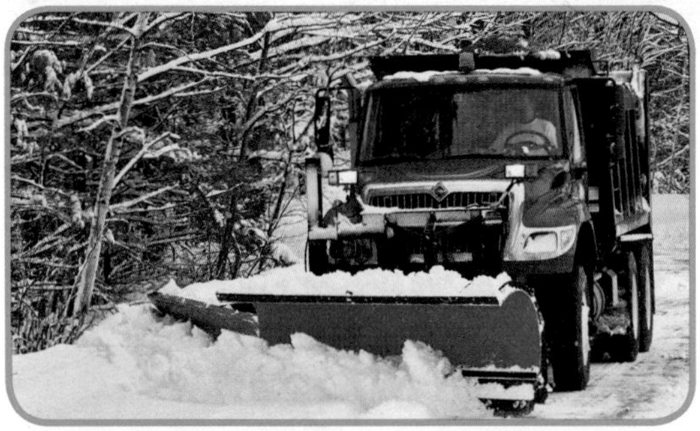

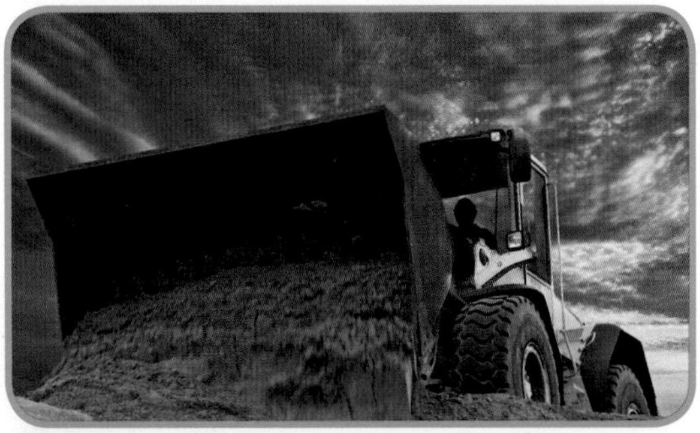

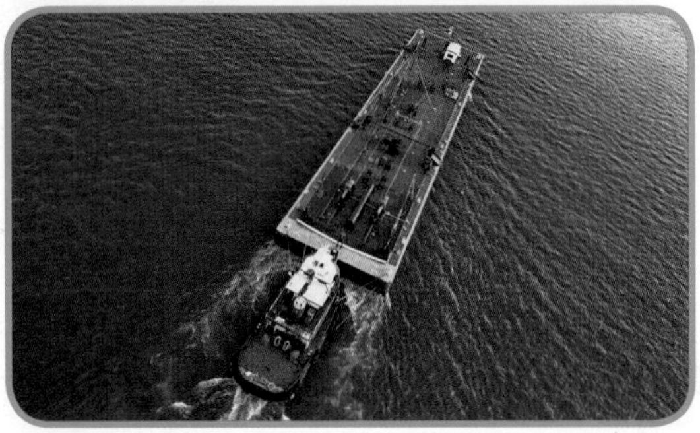

Pull

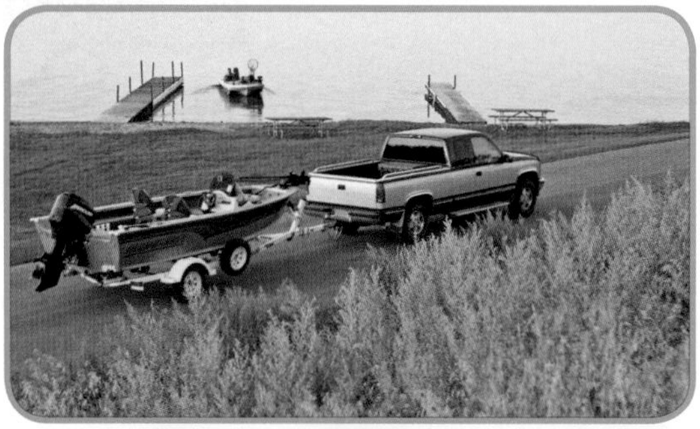

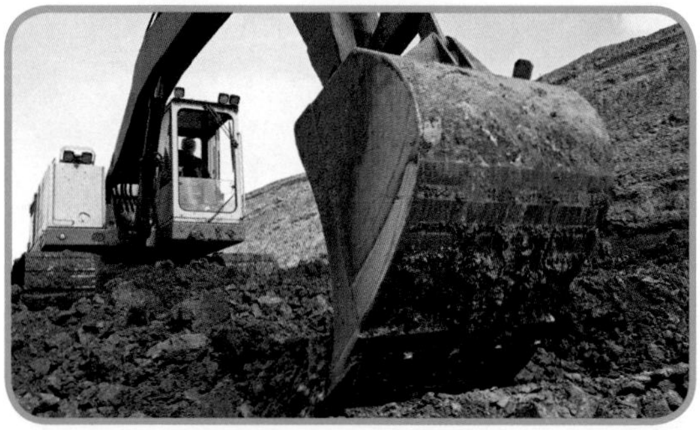

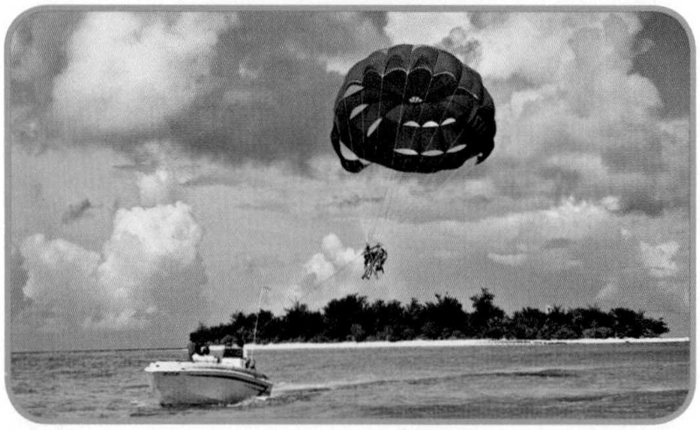